TECHNICAL
REPORT

Valuing Programmed Depot Maintenance Speed

An Analysis of F-15 PDM

Edward G. Keating, Elvira N. Loredo

Prepared for the United States Air Force

The research described in this report was sponsored by the United States Air Force under Contract FA7014-06-C-0001. Further information may be obtained from the Strategic Planning Division, Directorate of Plans, Hq USAF.

Library of Congress Cataloging-in-Publication Data

Keating, Edward G. (Edward Geoffrey), 1965–
Valuing programmed depot maintenance speed : an analysis of F–15 PDM / Edward G. Keating, Elvira N. Loredo.
p. cm.
Includes bibliographical references.
ISBN-13: 978-0-8330-3968-2 (pbk. : alk. paper)
1. Eagle (Jet fighter plane)—Maintenance and repair—Costs—Evaluation. I. Loredo, Elvira N. II. Title.

UG1242.F5K43 2006
358.4'383—dc22

2006028059

The RAND Corporation is a nonprofit research organization providing objective analysis and effective solutions that address the challenges facing the public and private sectors around the world. RAND's publications do not necessarily reflect the opinions of its research clients and sponsors.

Published 2006 by the RAND Corporation
1776 Main Street, P.O. Box 2138, Santa Monica, CA 90407-2138
1200 South Hayes Street, Arlington, VA 22202-5050
4570 Fifth Avenue, Suite 600, Pittsburgh, PA 15213-2665
RAND URL: http://www.rand.org/
To order RAND documents or to obtain additional information, contact
Distribution Services: Telephone: (310) 451-7002;
Fax: (310) 451-6915; Email: order@rand.org

Preface

Lt Gen Donald J. Wetekam, Deputy Chief of Staff for Logistics, Installations and Mission Support, Headquarters U.S. Air Force, and Maj Gen Arthur B. Morrill III, Director of Logistics, Headquarters Air Force Materiel Command, Wright-Patterson Air Force Base,[1] asked the RAND Corporation to develop a series of analyses and models to be used as vehicles for understanding the effects of changes in U.S. Air Force programs on operational capabilities.

As an initial case study, RAND evaluated the F-15 programmed depot maintenance (PDM) process as it occurs at the Warner Robins (WR) Air Logistics Center (ALC) at Robins Air Force Base in central Georgia. RAND studied the recent history of F-15 PDM at WR, including WR's recent implementation of "lean" approaches.

This report focuses on the issue of PDM speed. If PDM is faster, operating commands will possess more aircraft. What valuation should be attached to accelerated PDM? We present a methodology to estimate such value. This type of calculation would be relevant if the Air Force had to decide whether to invest funds to expedite PDM or whether to save funds through slower PDM.

RAND Project AIR FORCE has previously investigated issues related to the Air Force depot system. The resulting publications include the following:

- *How Should the U.S. Air Force Depot Maintenance Activity Group Be Funded?* Edward G. Keating and Frank Camm (MR-1487-AF). This monograph examines how Air Force Materiel Command depot-level expenditures relate to operating command activity levels. In it, the authors note a general lack of correlation between depot-level expenditures and fleet flying hours.
- *Aging Aircraft: USAF Workload and Material Consumption Life Cycle Patterns*, Raymond A. Pyles (MR-1641-AF). This monograph examines aging aircraft issues and potential future increases in PDM hours as aircraft age. Maintenance workloads and material consumption generally exhibited late-life growth as aircraft aged, but the rate of that growth depended on both the aircraft's flyaway cost and the workload category. Depot-level expenditures appeared to be the workload category most vulnerable to age-related increases.

[1] General Morrill was the Director of Resource Integration, Deputy Chief of Staff for Installations and Logistics, when he sponsored this research.

- *Aging Aircraft Repair-Replacement Decisions with Depot-Level Capacity as a Policy Choice Variable*, Edward G. Keating, Don Snyder, et al. (MG-241-AF). This monograph suggests that it might be appropriate to increase depot-level capacity to get highly valued aircraft through PDM more quickly. The authors evaluate the feasibility of either modifying or retiring the C-5A fleet and extend their modeling approach to evaluate prospective investment in additional depot-level capacity.

The research reported here was sponsored by the Deputy Chief of Staff for Logistics, Installations and Mission Support, U.S. Air Force (AF/A4/7), and the Director of Resource Integration, Deputy Chief of Staff for Logistics, Installations and Mission Support, U.S. Air Force (AF/A4P), and conducted within the Resource Management Program of RAND Project AIR FORCE. The work was performed as part of a fiscal year 2005 project titled Capability-Based Programming.

This report is intended to be of interest to Air Force and other Department of Defense maintenance and financial personnel.

RAND Project AIR FORCE

RAND Project AIR FORCE (PAF), a division of the RAND Corporation, is the U.S. Air Force's federally funded research and development center for studies and analyses. PAF provides the Air Force with independent analyses of policy alternatives affecting the development, employment, combat readiness, and support of current and future aerospace forces. Research is conducted in four programs: Aerospace Force Development; Manpower, Personnel, and Training; Resource Management; and Strategy and Doctrine.

Additional information about PAF is available on our Web site at http://www.rand.org/paf.

Contents

APPENDIXES

Figures

Tables

Summary

Every day (or hour) that a commercial airline operates an aircraft, it expects to generate a level of profit. Such a profit-per-day metric can then be used to assess the premium an airline would be willing to pay to get an aircraft through depot-level maintenance more quickly.

The U.S. Air Force lacks a profit metric for its aircraft. Yet, it faces cost-benefit calculations in its depot maintenance practices. Would it be worth investing $50,000 to expedite by a month an aircraft's PDM visit? How about $500,000?

This report presents a new methodology to calculate the value of expediting PDM. We use the fact that the Air Force has chosen to pay for intermittent PDM visits to estimate a defensible lower bound on what expedited PDM would be worth. We use F-15 data to illustrate our methodology.

The F-15 and Its Programmed Depot Maintenance

The F-15 is an all-weather, extremely maneuverable tactical fighter designed to permit the Air Force to gain and maintain superiority in aerial combat. F-15s receive PDM at the Warner Robins Air Logistics Center at Robins Air Force Base in central Georgia.

F-15s are generally on a six-year PDM cycle, i.e., they return to PDM six years after they leave. We assume that an F-15 stays in the fleet for 30 years, so we expect an aircraft to make four visits to PDM over its lifetime. Over the last six years, WR has produced 100–110 F-15 PDMs annually. In fiscal year (FY) 2005, the average duration of a completed F-15 PDM visit was about 130 days. (See pp. 6–8.)

A Simple Valuation of Expedited PDM

Our model supposes there must be enough net benefit (total benefit above incremental cost) after completion of a PDM visit to justify the cost of PDM. Fiscal year 2005 Air Force Total Ownership Cost system data suggest that a typical F-15 PDM visit during that year cost about $3.2 million. (See p. 11.)

There are different aircraft valuation curves consistent with a PDM visit being worthwhile. Assuming that net valuation does not increase as an aircraft ages, the most conservative valuation curve (generating the lowest value of expedited PDM) is a horizontal line.

With a horizontal valuation line, we estimate expediting an F-15's last PDM visit by one month would be worth about $60,000. A horizontal valuation line also implies that it is preferable to expedite an older, rather than newer, aircraft's PDM visit. (See pp. 13–14.)

Valuing F-15 PDM Speed with Declining Aircraft Valuation

We think aircraft tend to be worth less (adjusting for inflation) as they age. As time passes, potential adversaries obtain new technology that may render an aircraft less effective. Additionally, the aircraft may have declining availability and/or rising maintenance costs with age.

Unfortunately, we do not observe aircraft valuation over time. We do, however, observe aircraft mission capability (MC) and full mission capability (FMC) rates. F-15C/D MC and FMC rates increased substantially in the early months of calendar year 2002, but have otherwise undergone a long-term decline. A declining mission capable rate as an aircraft ages is consistent with declining aircraft valuation. Declining mission capability may cause declining valuation or it may be a symptom of declining valuation. (See pp. 15–18.)

We incorporated declining aircraft valuation into our PDM acceleration valuation calculation. With a 1.35-percent annual valuation decline rate (consistent with the observed F-15C/D FMC rate of decline), expediting an F-15's last PDM visit is estimated to be worth at least $74,366 (up from $60,639 with constant valuation). More pronouncedly, our estimates of the value of accelerating earlier PDM visits for newer aircraft increase markedly, e.g., accelerating a newer F-15's first PDM visit is worth more than $180,000. Acceleration values are greater using a 1.7-percent annual valuation decline rate consistent with the observed F-15C/D MC rate of decline. (See pp. 18–22.)

We find it reasonable and intuitive that expediting a newer aircraft's PDM visit is more valuable than expediting an older aircraft's visit.

Robustness Explorations

Previous RAND research (see, for example, Pyles, 2003) has documented aging aircraft effects, such as rising maintenance costs as aircraft age.

Using plausible, though purely illustrative, aging aircraft maintenance cost growth parameters, we repeated our estimation of PDM acceleration valuation.

Incorporation of aging aircraft maintenance cost effects consistently raises our estimated value of PDM acceleration. In particular, when the fourth and final PDM visit is more expensive, aircraft valuation throughout the life cycle must be greater, assuming that undertaking the last PDM visit was appropriate. (See pp. 23–25.)

We also explored an additional constraint that an aircraft's life-cycle net benefits must equal or exceed its life-cycle costs, including acquisition costs.

If aircraft valuation is assumed to be level over an aircraft's life span, imposition of this additional constraint is very important and drives up the implied valuation of expedited PDM markedly. If, however, aircraft valuation is assumed to decline over time, imposing this additional acquisition cost constraint makes little (1.35-percent valuation decline case) or no (1.7-percent valuation decline case) difference in our estimates of the value of accelerated PDM. (See pp. 26–29.)

We also explored a structure in which aircraft valuation jumps after PDM visits. Such jumps reduce the estimated value of accelerating earlier PDM visits but have no effect on the estimated value of accelerating the last PDM visit. (See pp. 29–31.)

Acknowledgments

We especially thank John Fisher and Chandra Thompson for their roles as our points of contact at Warner Robins. We also thank Goran Bencun, Rena Britt, Steve Brooks, Lt Col Alex Cruz-Martinez, Doug Daniels, Ellen Griffith, Dale Halligan, Norma Jacobs, Alan Mathis, Sergeant Kennita Mathis, Jeff Owens, Lorie Snipes, John Stone, and Ken Winslette of WR.

We received helpful insight for our work from Col Stephen Sheehy of AF/A8E, Lt Col Lawrence Audet of AF/A4PE, and Timothy Groseclose of the University of California, Los Angeles. We received constructive reviews of this document from our colleagues Frank Camm and John Schank. We also thank our colleagues Susan Bowen, Tony Bower, Cynthia Cook, Herman (Les) Dishman, Greg Hildebrandt, Kent A. Hill, Richard J. Hillestad, Robert Leonard, Adam Resnick, Charles Robert Roll, Jr., Roberta M. Shanman, Leslie Thornton, Eric Unger, and Mark Wang for assistance on this research. Jane D. Siegel helped prepare this document, and Lauren Skrabala edited it.

An earlier version of this research was briefed at the Western Economic Association annual conference in San Francisco, Calif., on July 5, 2005. We appreciate the comments of our discussant, Francois Melese of the Naval Postgraduate School.

Of course, the authors alone are responsible for errors that remain in the document.

Abbreviations

AF/A4/7	Deputy Chief of Staff for Logistics, Installations and Mission Support, U.S. Air Force
AF/A4P	Director of Resource Integration, Deputy Chief of Staff for Logistics, Installations and Mission Support, U.S. Air Force
AFTOC	Air Force total ownership cost
ALC	Air Logistics Center
FMC	full mission capability
FY	fiscal year
IAP	international airport
LN	natural logarithm
MC	mission capability
PAF	Project AIR FORCE
PDM	programmed depot maintenance
RAF	Royal Air Force
REMIS	Reliability and Maintainability Information System
WR	Warner Robins

CHAPTER ONE

Introduction

The U.S. Air Force asked the RAND Corporation to study capability-based programming. The long-term goal is to develop a series of analyses and models to understand the effects of changes in Air Force programs on operational capabilities. If funding is increased, how might capability be improved? If funding is cut, how might capability be degraded?

Depot maintenance funding influences capability. Aircraft enter programmed depot maintenance (PDM) on a regular schedule. The level of resources devoted to PDM influences both how much work is done in PDM (i.e., how much more reliable or capable aircraft are after leaving PDM) and the duration of PDM. Other things equal, we expect a better-funded process to run more quickly, e.g., there are fewer queues within the depot and more spare parts available.

In this report, we focus on the issue of PDM speed. When PDM is lengthy, more aircraft are tied up in PDM at any given point in time; fewer aircraft are available to operating commands. It would be desirable to expedite PDM: Aircraft would spend a greater fraction of their lives in the possession of operating commands and available for usage, if required. In this report, we present a new methodology to estimate the value of accelerated PDM.

For a commercial airline, calculating the value of expedited maintenance is (relatively) straightforward: A commercial airliner is expected to generate a certain amount of profit each day (or hour) it operates. Lost profit forms a benchmark for the value of accelerating commercial airliner PDM (which airlines term *D checks*). Not surprisingly, if demand for commercial aviation is soft, an airline will be less willing to devote resources to expediting aircraft maintenance.

Military aircraft lack such a profit metric. Yet, some valuation of military aircraft in operating command possession is necessary if the Air Force is to assess the desirability of investing resources in expediting PDM (or saving money by slowing PDM). The methodology presented in this report is intended to inform depot-level cost-benefit analysis. Would it be worth investing $50,000 to expedite an aircraft's PDM by a month? How about $500,000?[1]

We focus on the F-15 fighter aircraft in this work. We chose the F-15 as an initial illustrative example with the agreement of the Air Force. The F-15 is a very valuable part of the Air

[1] In this report, we assume that it costs something to expedite PDM. If PDM could be relatively costlessly expedited (e.g., through a process reorganization or changing labor practices), we assume that such reforms would have already been implemented. Any improvement not yet implemented must have a cost (or else it would already be in place, we assume).

Force's fleet, so its inclusion in this study is inherently important. More broadly, the issues and tradeoffs relevant to the F-15 may apply to other aircraft as well. So, while this analysis focuses on the F-15, we believe the methodology we present is more widely applicable.

The methodology builds on revealed preferences. In particular, throughout an aircraft's life, we observe the Air Force making choices, e.g., to put an aircraft through PDM.[2] We logically infer, therefore, that an aircraft must have sufficient expected net benefits (total benefits less incremental operating costs) after PDM to justify the cost of that PDM visit. (If this were not so, the Air Force would have been better off retiring the aircraft rather than undertaking PDM.) This approach is similar to the method of Reinertsen et al. (2002), which assumes that the benefits of a system are at least equal to its costs.

This thought experiment can be worked back all the way to the aircraft's initial purchase. When a given aircraft was initially acquired, the Air Force must have projected a stream of total benefits from the aircraft that equaled or exceeded the stream of costs generated by the aircraft. In Chapter Five of this report, we discuss the consequences of imposing the constraint that life-cycle benefits equal or exceed life-cycle costs. There are several concerns with this constraint. First, the aircraft's purchaser may have misestimated the stream of future costs (many of which were decades away at the time of aircraft's acquisition). Second, the vast majority of the F-15 fleet, for example, was acquired during the Cold War. A Cold War–era decisionmaker doubtlessly had a very different perception of the aircraft's operating environment than proved to be the case.

By contrast, the decision to put an F-15 through PDM is a much simpler one to which the decisionmaker brings much more current knowledge. We assume, for exogenous safety reasons, that an F-15 that has been operated for six years (72 months) must either undergo a PDM visit or retire. If PDM is undertaken, an up-front fee is paid. (Using fiscal year 2005 Air Force Total Ownership Cost system data discussed below, we estimate this fee to be about $3.2 million.) The refurbished aircraft then returns to its operating command for another six years of operation (measured from the date the aircraft exits PDM). We also assume that an F-15 must be retired after 30 years of ownership. (As a consequence, the stream of benefits after the fourth and last PDM visit, around age 25, is briefer than earlier visits' streams.)

Assuming that each of the decisions to put F-15s through PDM was correct, the following statements must hold true:

- Net benefits after the fourth PDM visit must equal or exceed the cost of the fourth PDM visit.
- Net benefits after the third PDM visit (including those after the fourth visit) must equal or exceed the cost of the third visit. (The cost of the fourth PDM visit is built into computation of net benefits after the third visit.)

[2] Another possibility, of course, is that the Air Force wanted to retire an aircraft but Congress instead decided the aircraft should be kept and put through PDM. For the sake of parsimony, we speak of "the Air Force making choices," but, quite properly, ultimate decisionmaking in the system lies with elected officials.

- Net benefits after the second PDM visit must equal or exceed the cost of the second visit.
- Net benefits after the first PDM visit must equal or exceed the cost of the first visit.

Once we have estimated a floor on an aircraft's net benefits (i.e., the minimum net benefit flow that satisfies these constraints), it is straightforward to compute a floor on what an extra month of operator availability from expedited PDM would be worth.

Our philosophy, throughout this report, is to estimate a defensible lower bound on what expedited PDM might be worth. Aircraft may be worth far more than what the Air Force pays for their PDM. But we restrict ourselves, instead, to the decisions we observe, e.g., to undertake PDM, and draw inferences as to what those decisions (assuming they are appropriate) minimally imply about aircraft valuation.

While this report focuses on the F-15, our conceptual approach is applicable to any aircraft or other vessel (such as a ship) that intermittently enters depot-level maintenance. The logic is always the same: Net benefits after the depot visit must equal or exceed the costs of that visit.

The next chapter provides information about the F-15 and its PDM program. Chapter Three presents a simple model of valuing expedited PDM. Chapter Four presents our main PDM-acceleration valuation results as we build in aircraft valuation that declines over time. Chapter Five assesses the robustness of our findings considering aging aircraft phenomena, an additional constraint that an aircraft's life-cycle net benefits cover acquisition costs, and possible jumps in aircraft valuation following PDM visits. Chapter Six concludes the report with different estimates of the value of expediting a PDM visit by one month. Appendix A presents more details about how our acceleration valuation calculations were undertaken in Microsoft® Excel®. In Appendix B, we discuss how the size of a new military aircraft fleet might be simultaneously determined with its PDM speed.

CHAPTER TWO

The F-15 and Its Programmed Depot Maintenance

The F-15 is an all-weather, extremely maneuverable, tactical fighter designed to permit the Air Force to gain and maintain superiority in aerial combat.[1] As shown in Table 2.1, there are five F-15 variants. The first four variants were designed for air-to-air combat, while the newest, most capable, and most expensive variant, the F-15E, combines air-to-air and air-to-ground attack capabilities. The data in Table 2.1 are from the Reliability and Maintainability Information System (REMIS)[2] and were current as of the end of September 2005.

All five variants have a distinctive double vertical tail design; Figure 2.1 is a photograph of an F-15.

Table 2.2 shows the assigned locations of the 722 F-15s as of the end of September 2005. Of those, 684 were at operating commands' installations and 38 were possessed by the depot system.

Table 2.1
F-15 Variants

Variant	Seats	Primary Mission	Number Operating	Acceptance Date of Oldest Aircraft	Acceptance Date of Newest Aircraft
A	1	Air-to-air combat	92	March 5, 1975	April 2, 1981
B	2	Air-to-air combat training	14	March 24, 1976	April 18, 1979
C	1	Air-to-air combat	338	May 24, 1979	October 20, 1989
D	2	Air-to-air combat training	54	June 22, 1979	August 11, 1987
E	2	Air-to-ground attack	224	March 11, 1987	September 28, 2004

SOURCE: Data from the Reliability and Maintainability Information System.

[1] For a fact sheet on the F-15, see U.S. Air Force (2005).

[2] REMIS tracks a variety of statistics Air Force–wide, including flying hours and landings by individual aircraft, as well as various categories of maintenance activities at the individual aircraft level. As shown in Table 2.1, REMIS also tracks when aircraft first enter the Air Force (acceptance dates).

Figure 2.1
An F-15

SOURCE: U.S. Air Force photo by Master Sgt. Val Gempis.
RAND *TR377-2.1*

The Warner Robins (WR) Air Logistics Center (ALC) at Robins Air Force Base in central Georgia provides PDM to all five F-15 variants. F-15s are typically on a six-year PDM cycle, i.e., six years after completion of PDM, they are due to return. The five variants are all handled on the same PDM line at WR, albeit with some procedural changes related to the aircrafts' configuration differences.

We assume that F-15s are operated until they are 30 years (360 months) old.[3] Hence, an aircraft is expected to have four PDM visits during its lifetime.

Figure 2.2 plots the duration (in calendar days) of F-15 PDMs completed at WR in fiscal years (FYs) 2000–2005. The horizontal axis is the date the PDM work was completed. WR provided these data.

[3] In this analysis, we stipulate that the aircraft will be retired after 30 years of service. In Keating and Dixon (2003), we presented a methodology to determine optimal retirement age. Our focus here is different, so we simply assume 30-year (360-month) retirement throughout.

Table 2.2
Assigned Locations of F-15s, End of September 2005

Location Type	Location	F-15 Class					Total
		A	B	C	D	E	
Operating command							
	Seymour Johnson AFB, N.C.	0	0	0	0	92	92
	RAF Lakenheath, UK[a]	0	0	22	2	54	78
	Tyndall AFB, Fla.	0	0	46	25	0	71
	Eglin AFB, Fla.	0	0	54	6	4	64
	Elmendorf AFB, Alaska	0	0	40	4	19	63
	Langley AFB, Va.	0	0	47	6	0	53
	Kadena Air Base, Japan	0	0	46	4	0	50
	Mountain Home AFB, Ida.	0	0	17	1	28	46
	Nellis AFB, Nev.	0	0	14	2	14	30
	Hickam AFB, Hawaii	18	2	0	0	0	20
	Jacksonville IAP Air Guard Station, Fla.	19	1	0	0	0	20
	Kingsley Field, Oreg.	0	6	11	3	0	20
	Lambert–St. Louis IAP Air Guard Station, Mo.	0	1	17	1	1	20
	New Orleans Air Reserve Station, La.	19	1	0	0	0	20
	Otis AFB, Mass.	17	1	0	0	0	18
	Portland AFB, Oreg.	16	2	0	0	0	18
	Holloman AFB, N.M.	0	0	1	0	0	1
Depot system							
	Warner Robins ALC, Ga.	3	0	20	0	12	35
	Kimhae Depot, South Korea[b]	0	0	3	0	0	3
Total		92	14	338	54	224	722

NOTES: RAF = Royal Air Force. IAP = International Airport.
[a] RAF Lakenheath is a U.S. Air Force–operated fighter base.
[b] Kimhae is a Korean Airlines–operated facility that provides PDM to F-15s based at Kadena Air Base, as well as to aircraft based at other Pacific Air Forces locations.

There has been some upward drift in the average PDM time over recent years, as shown in Table 2.3. The fiscal years presented are those in which the PDM work ended. In some cases, the PDM visit commenced in the preceding fiscal year.

Figure 2.2
WR F-15 PDM Durations

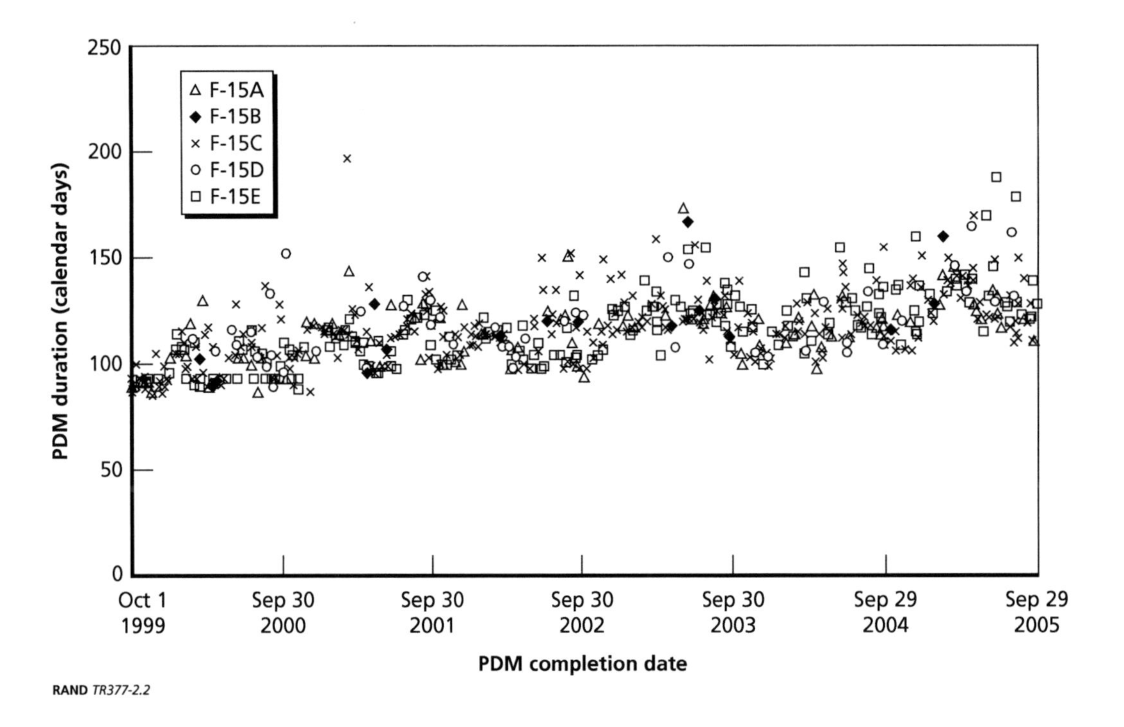

RAND *TR377-2.2*

In recent years, WR has adopted a "lean" approach to its F-15 PDM process. One possible interpretation of Table 2.3 is that the lean innovations have not yet had a great effect on average PDM speed. Another possibility is that the process would have slowed more markedly due to aging aircraft problems and/or loss of experienced labor, but the transition to a lean approach has lessened the impact of such challenges.

Table 2.3
Completed F-15 PDM Durations

	Fiscal Year					
	2000	2001	2002	2003	2004	2005
Number completed	116	108	108	102	103	101
Mean days	99.5	112.7	111.5	123.1	117.5	130.3
Standard deviation	11.2	14.8	11.5	13.7	12.2	15.7
Median days	96.5	112.5	112	122	116	128
75th percentile	106.3	119.3	117	127	125	138
95th percentile	119	134	133	153	142	162

SOURCE: Data supplied by WR ALC.

If F-15 PDM were expedited, the most immediate effect would be an extra month of aircraft availability for its operator. Consider a hypothetical F-15 that enters PDM on January 1, 2006. If its PDM visit lasts four months, it would return to its operating command on May 1, 2006. By contrast, if the PDM process were expedited by a month, the operator would receive its aircraft on April 1, 2006. Thus, April 2006 would be a "gained month" of operator availability that would be accrued through expedited PDM.

Unless the PDM visit commencing on January 1, 2006, was an aircraft's last before retirement, there would be an additional effect to expedited PDM. Aircraft are to return to PDM six years after departing, so the expedited aircraft would be due back for PDM a month sooner than would otherwise be the case (April 1, 2012, rather than May 1, 2012, in this example). Hence, future PDM visit(s) and their associated costs would be shifted forward if PDM were expedited.

In the next chapter, we present a simple methodology to value expedited PDM. Our methodology accounts for both the immediate extra-month effect of expedited PDM as well as the consequences of shifting forward an aircraft's future PDM visits.

CHAPTER THREE

A Simple Valuation of Expedited PDM

This report aspires to provide useful input for depot-level cost-benefit analysis. How much benefit would accrue from expedited F-15 PDM? For example, would it be worthwhile to invest $100,000 in a piece of equipment that expedites an aircraft's PDM visit by 30 days?

In this chapter, we present the simplest and most conservative version of our PDM speed valuation technique. We calculate a lower bound on the worth of expediting F-15 PDM by one month. (In subsequent chapters, we enhance our model, resulting in increases in our estimated value of expedited PDM.)

As discussed previously, our model supposes that there must be enough net benefit (total benefit above incremental cost) after completion of a PDM visit to justify the cost of PDM. (Otherwise, the aircraft should have been retired rather than going through PDM.)

With a six-year PDM cycle, a 30-year lifespan, and assumed 120-day PDM visits, an F-15 would be possessed by its operating command in months 1 through 72 (the first six years), months 77 through 148 (the next six years after the four-month PDM), months 153 through 224, months 229 through 300, and months 305 through 360.

We do not have different estimates of the costs of the four PDM visits. The Air Force total ownership cost (AFTOC)[1] system, however, indicates that FY 2005 F-15 depot maintenance expenditures totaled about $358 million. In FY 2005, WR completed PDM on 101 F-15 aircraft plus unscheduled depot-level maintenance on two aircraft, while Kimhae completed 10 F-15 PDMs. Hence, the average expenditure per completed F-15 aircraft was about $3.2 million in FY 2005 ($358 million divided by 113). So, we assume that each PDM visits costs about $3.2 million.

Let us focus, for a moment, on the fourth and final PDM visit. If its cost equals $3.2 million, the discounted sum of net benefits from months 305–360 must at least equal $3.2 million:

$$\sum_{m=305}^{360} \frac{(B_m - C_m)}{(1+d)^{\left(\frac{m-301}{12}+\frac{1}{24}\right)}} \geq \$3.2\text{M},$$

[1] The AFTOC data system tracks annual expenditures in various categories attributed to different weapon systems. Depot maintenance is one of the AFTOC categories. AFTOC also breaks up expenditures across the five F-15 variants, though a number of allocations are required to separate, for instance, F-15C from F-15D expenditures. We use F-15–wide data in this analysis with the assumption that such data are more reliable than data broken up by F-15 variant.

where B_m is the constant (FY 2005) dollar benefit in month m; C_m is the constant dollar incremental cost in month m; and d is the annual discount rate. The Office of Management and Budget prescribes a 2006 annual real interest rate of 3.0 percent.[2] We assume costs and benefits accrue in the middle of each month so the net benefit in month 305, for instance, is 4.5 months away measured from the beginning of month 301, hence the complicated exponent on $(1+d)$.

In this chapter, we assume real net benefit $(B_m - C_m)$ does not increase as an aircraft ages. It follows that net surplus must be non-negative in months 305–360 (or else aircraft retirement should have come sooner than month 360). Hence, $B_m \geq C_m \forall m \in [305,360]$.

Simply knowing that months 305–360 have a discounted total net surplus of $3.2 million and none of the months has negative surplus does not tightly identify the prospective net surplus in month 304, i.e, how much an extra month of operating command possession would be worth. Figure 3.1 presents two different cases in which the cumulative surplus for months 305–360 sums to $3.2 million.

In the triangular valuation case in Figure 3.1, net surplus decreases linearly until month 360, when it is zero. If we trace the triangular valuation line in Figure 3.1 back to month 304, one projects a net surplus of $121,704 in that month or $120,660 measured from the beginning of the PDM visit in month 301.

Figure 3.1
Different Net Surplus Cases Consistent with the Fourth PDM Being Worthwhile

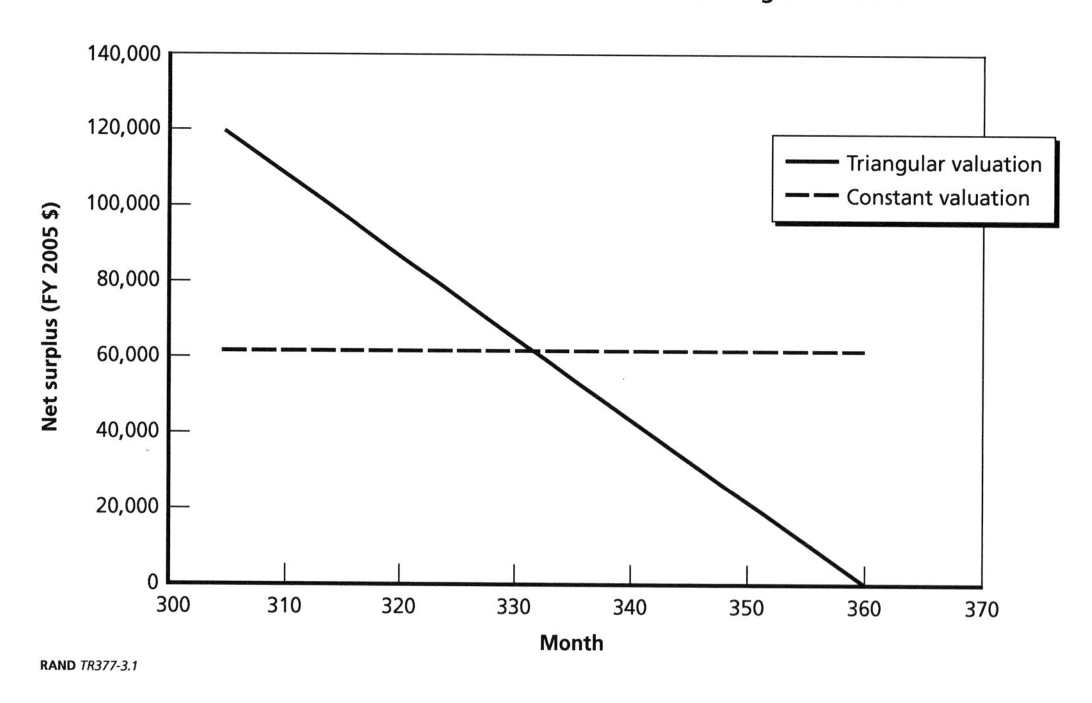

RAND *TR377-3.1*

[2] See Office of Management and Budget (2006).

With constant valuation, by contrast, monthly net surplus is steady at $61,164, so a hypothetical month 304 would also have that valuation (which would be worth $60,639 at the commencement of the fourth PDM visit at the beginning of month 301). Assuming the net surplus $(B_m - C_m)$ line cannot rise (i.e., aircraft do not get more valuable as they get older), $60,639 is the lowest possible value of expediting the fourth and last PDM visit by one month, consistent with that PDM visit being worth undertaking.

We can use the same technique to value accelerating an earlier PDM visit. If net benefits after the fourth PDM visit at least equal $3.2 million, a minimal condition for the third PDM visit to be worthwhile is

$$\sum_{m=229}^{300} \frac{(B_m - C_m)}{(1+d)^{(\frac{m-225}{12}+\frac{1}{24})}} \geq \$3.2\text{M},$$

and comparably that

$$\sum_{m=153}^{224} \frac{(B_m - C_m)}{(1+d)^{(\frac{m-149}{12}+\frac{1}{24})}} \geq \$3.2\text{M}$$

(the second PDM visit is worthwhile), and

$$\sum_{m=77}^{148} \frac{(B_m - C_m)}{(1+d)^{(\frac{m-73}{12}+\frac{1}{24})}} \geq \$3.2\text{M}$$

(the first is worthwhile).

If we assume, however, that $B_m - C_m$ is nonincreasing in m (ignoring the $B_m = 0$ months during PDM visits), none of these constraints is binding if the "fourth PDM is worthwhile" constraint is fulfilled. We assume only 56 months of operation after the fourth PDM visit, so any net valuation high enough to make the fourth PDM worthwhile makes the earlier PDM visits (with 72 months of operation until the next PDM visit) worthwhile.

Thus, assuming that real net benefit does not increase with the age of an aircraft, a horizontal net surplus line with a monthly net surplus of $61,164 is the lowest possible net surplus curve consistent with all our constraints (all four PDM visits are worthwhile; real net benefit is not increasing, ignoring the valuation jumps from zero following PDM visits).

There is a curiosity in the use of a horizontal constant valuation line, however. As noted, we estimate that expediting the last PDM visit by one month would be worth $60,639 (measured from the start of that visit). With our model, expediting the third PDM visit by a month would be worth less: $53,661. As with expediting the last PDM visit, one immediately gains a month of availability (April 2006, using the example in Chapter Two). But an additional disadvantage when the third PDM visit is expedited by a month is that the $3.2 million bill for the fourth and last PDM visit shifts forward by a month. One also loses availability in April 2012 while gaining it in August 2012 (assuming PDM is back to a 120-day calendar in 2012).

For the second and first PDM visits, there are two and three future "month shifts," respectively, the result of expediting PDM by one month. With a horizontal valuation line, none of these shifts is favorable. One repeatedly shifts forward $3.2 million PDM bills. Hence, with our constant valuation net surplus line over the aircraft's life, we find that expediting a new F-15's first PDM visit is least valuable ($43,077), while expediting a second PDM visit by a month would be worth more than expediting a first visit but less than expediting a third ($47,875).

We do not believe it to be reasonable that it is most valuable to expedite an old aircraft's last PDM visit. Instead, our intuition is that aircraft become less valuable as they age and the best aircraft on which to focus resources in terms of expediting PDM are the newest ones.

In the next chapter, we present an enhanced PDM valuation approach that appropriately considers declining aircraft valuation.

CHAPTER FOUR

Valuing F-15 PDM Speed with Declining Aircraft Valuation

Intuitively, we suspect that an aircraft is most valuable when it is new: Its systems are closer to the technological frontier; the aircraft has not suffered wear and tear; there are no aging aircraft problems. As time passes, potential adversaries obtain new technology that may render an aircraft less effective. Additionally, the aircraft itself may have declining availability and/or rising maintenance costs. Barring some exogenous increase in demand for the aircraft, valuation should be expected to decline as aircraft age (though, by construction, valuation jumps up after $B_m = 0$ periods of PDM).

The rate at which aircraft valuation declines is an important parameter in a PDM acceleration calculation. If valuation declines quickly (e.g., the triangular valuation case in Figure 3.1), one derives a much greater value from accelerated PDM than in the case of slow valuation decline (or, in the limit, constant valuation in Figure 3.1).

Unfortunately, the "right" valuation decline rate is unknowable because some of its inputs are unknowable (e.g., the rate at which potential adversary technology is gaining on the F-15). In this report, our more modest goal is to estimate a defensible lower bound on the rate at which the Air Force's valuation of the F-15 is declining and, consequently, a defensible lower bound on what expediting F-15 PDM might be worth. (The more rapid the valuation decline, the greater the valuation of PDM acceleration, other things being equal.)

To gain some insight on how F-15 valuation may be evolving over time, we next examine recent patterns in F-15 mission capability rates.

F-15 Mission Capability Rates

An aircraft is said to be have full mission capability (FMC) if it can perform all its missions. An aircraft is partially mission capable if it can fly safely with appropriate restrictions, but some of its capabilities are not functioning correctly. For example, an aircraft might still fly during daylight hours even if its night-vision equipment is not working. An aircraft is mission capable if it is either fully or partially mission capable. (The mission capability [MC] rate is, by construction, always greater than or equal to the FMC rate.)

In Figure 4.1, we display the monthly average mission capable and fully mission capable rates for F-15Cs and F-15Ds possessed by operating commands (excluding those aircraft held by Warner Robins and Kimhae) between October 1994 (the start of FY 1995) and September 2005 (the end of FY 2005).[1] These data are from REMIS.

As Figure 4.1 shows, F-15C/D mission capable and fully mission capable rates are nearly the same. (By contrast, Keating and Dixon [2003] showed it was common for KC-135s to be mission capable, but not fully mission capable).

A vertical axis has been placed in Figure 4.1 at September 11, 2001. In the months following the terrorist attacks (especially in the first half of calendar year 2002), F-15C/D mission capable and fully mission capable rates increased noticeably. Table 4.1 shows these monthly rates for August 2001–July 2002.

Figure 4.1
F-15C/D Monthly Mission Capability Rates

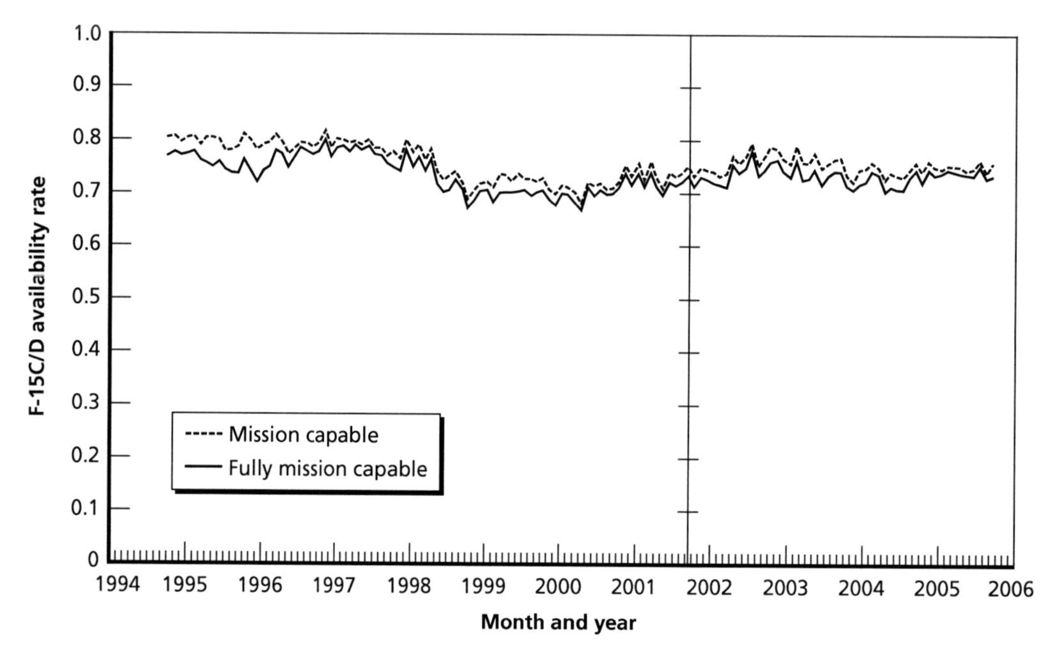

SOURCE: Data from the Reliability and Maintainability Information System.
RAND *TR377-4.1*

[1] We focused our analysis on the F-15C/D fleet because a sizable number of F-15A/Bs were retired during this period and new F-15Es were being added. Except for minimal attrition, the F-15C/D fleet was static, size-wise, between FY 1995 and FY 2005.

Table 4.1
F-15C/D Mission Capable and Fully Mission Capable Rates

Month	Mission Capable Rate (%)	Fully Mission Capable Rate (%)
August 2001	73.8	72.1
September 2001	75.0	73.2
October 2001	73.0	71.2
November 2001	74.8	73.1
December 2001	74.2	72.6
January 2002	73.8	71.8
February 2002	73.0	71.5
March 2002	73.6	71.0
April 2002	77.0	75.3
May 2002	75.3	73.8
June 2002	76.8	74.6
July 2002	79.3	78.0

Using the 11 years of monthly data, we ran regressions of the natural logarithm of the mission capable and fully mission capable rates on the monthly average ages of the F-15C/D fleet plus a dichotomous variable indicating 1 for months starting October 2001 (the first full month after the attacks). Table 4.2 shows our regression results.

The two regressions tell similar stories. In both cases, there is a highly statistically significant downward trend in the mission capability rate as aircraft age, interrupted by a one-time increase in the rate after September 11, 2001, then drifting down again. Figures 4.2 and 4.3 display these regression results.

Figure 4.2 depicts a 1.7-percent rate of decline in the mission capable rate for each year of aircraft age. Figure 4.3 depicts a 1.35-percent rate of decline in the fully mission capable rate for each year of aircraft age.

A declining mission capable rate as an aircraft ages is consistent with declining aircraft valuation. Causality could run in either or both directions. An aircraft that is less frequently mission capable is likely to be worth less to the Air Force. Alternatively or additionally, an aircraft that is worth less to the Air Force may receive reduced budgetary support, thereby reducing its mission capability rate. Either way, Table 4.2 and Figures 4.2 and 4.3 are consistent with the Air Force's declining valuation of the F-15 as the system ages.[2]

We next enhance our PDM acceleration valuation calculation to incorporate a declining valuation rate as aircraft age.

---

[2] In accord with this finding, Younossi et al. (2002) observe that Navy E-2C average flight hours per month decline with age. The better capability of new aircraft is one possible explanation; better reliability in new aircraft is another.

Table 4.2
F-15C/D Mission Capable and Fully Mission Capable Rate Regressions

Dependent Variable	LN(MC Rate)	LN(FMC Rate)
Intercept	−0.0234	−0.1092
Intercept standard error	0.0210	0.0237
Intercept *t*-statistic	–1.1141	–4.6055
Intercept *p*-value	0.0267	0.0000
Average age	−0.0172	−0.0135
Average age standard error	0.0014	0.0016
Average age *t*-statistic	–12.3202	–8.5665
Average age *p*-value	0.0000	0.0000
Post-9/11/01	0.0885	0.0727
Post-9/11/01 standard error	0.0093	0.0105
Post-9/11/01 *t*-statistic	9.5382	6.9408
Post-9/11/01 *p*-value	0.0000	0.0000
R-squared	0.5432	0.3629
Regression F	76.7087	36.7475
Observations	132	132

NOTE: LN = natural logarithm.

Incorporating a Declining Valuation Rate

In Figure 3.1 in Chapter Three, the triangular valuation portrayal showed the aircraft's net surplus declining linearly. In this section, we consider a decline in the aircraft's total valuation using the estimated rates of capability decline (1.35 percent for full mission capability and 1.7 percent for mission capability).

To consider total, not net, valuation, we must consider the F-15's expenditure structure more thoroughly. Table 4.3 presents more complete data from AFTOC on FY 2005 F-15 expenditures.

Figure 4.2
F-15C/D Mission Capable Rate–Age Regression

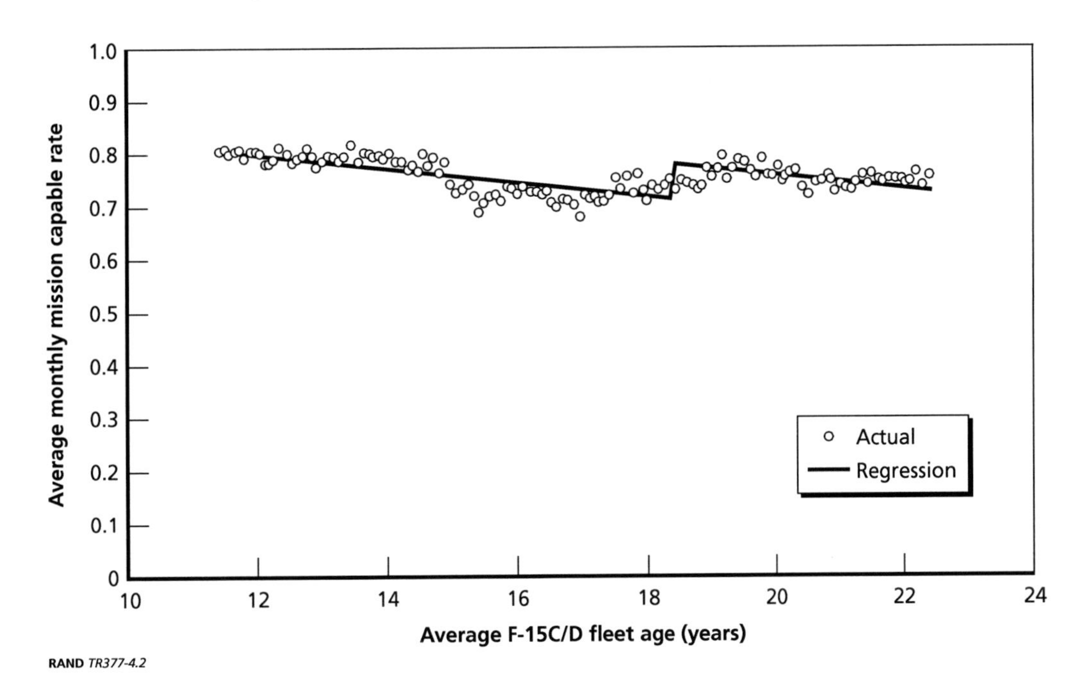

RAND *TR377-4.2*

Figure 4.3
F-15C/D Fully Mission Capable Rate–Age Regression

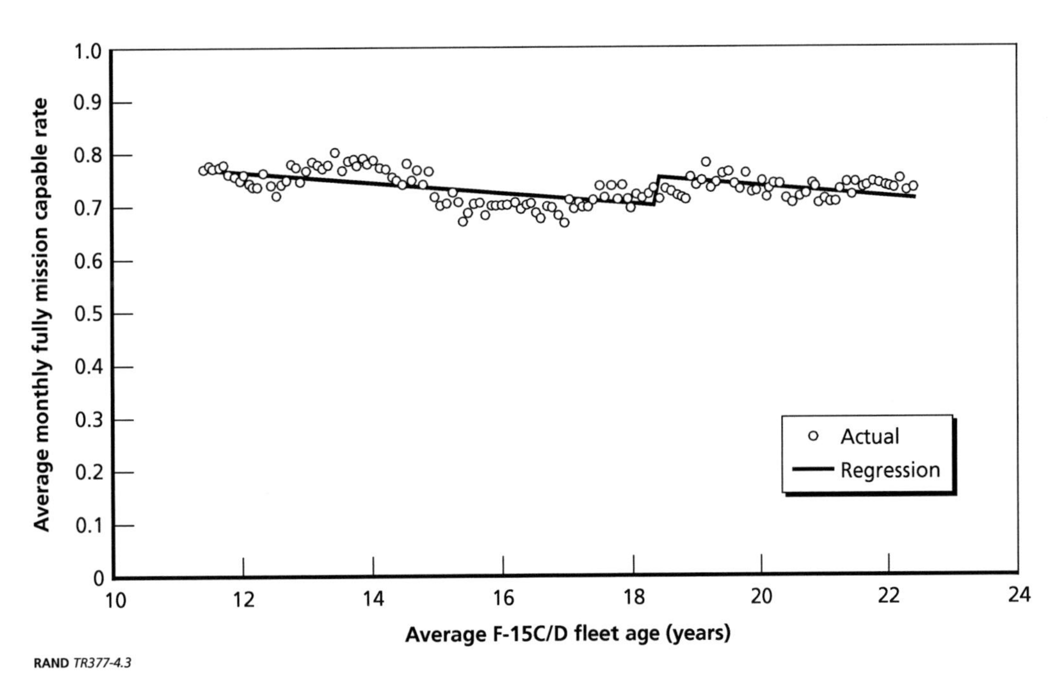

RAND *TR377-4.3*

On average, in FY 2005, about 685 F-15s were not in the depot system at any point in time, so ordinary operations expenditures per non-depot aircraft were about $4.5 million per year ($3095.8 million divided by 685) or about $376,000 per month. We assume, therefore, an F-15 not undergoing PDM incurs a marginal cost (C_m) of about $376,000 per month. Monthly valuation of the aircraft (B_m) must therefore equal or exceed $376,000 through month 360 (or else aircraft retirement should come sooner).

As was true in Figure 3.1, there are multiple shapes of aircraft valuation consistent with monthly valuation equal to or exceeding $376,000 and the fourth PDM being worthwhile. Based on the observed pattern of declining mission capability and full mission capability rates as aircraft age, we do not think constant valuation is appropriate. Instead, in Figure 4.4, we show valuation curves with 1.35- and 1.7-percent annual valuation decline (while satisfying the constraint that the last PDM is worth undertaking). These curves imply that accelerating the last PDM by one month would be worth $74,366 (1.35-percent case) or $77,969 (1.7-percent case), up from $60,639 in the flat valuation case. (Appendix A discusses how we undertook these calculations using Microsoft Excel.)

A declining, rather than flat, aircraft valuation curve is especially important for earlier PDM visits. In Figure 4.5, we extend the 1.35- and 1.7-percent valuation decrease cases throughout the life span of the aircraft. The four marginal cost spikes represent the four scheduled PDM visits. Aircraft valuation is assumed to be zero during PDM visits.

With these declining valuation curves, it is most valuable to expedite a new aircraft's PDM visit. With the 1.35-percent valuation decline rate, we estimate that accelerating the first PDM visit by a month would be worth $181,639, while the same acceleration would be worth $224,543 with a 1.7-percent annual decline rate (versus only $43,077 if aircraft valuation is constant through its life). As in Chapter Three, one is shifting forward $3.2 million in PDM costs, but the disadvantages of doing so are offset by the advantages of getting extra availability when aircraft are newer and more valuable. In Figure 4.6, we plot the estimated values of accelerating each of the four PDM visits by a month as a function of the annual valuation decline rate. The points in the middle of the four lines are the 1.35- and 1.7-percent valuation decline values.

We find it plausible and intuitive that expediting a newer aircraft's PDM visit is more valuable than expediting an older aircraft's visit. As discussed in Chapter Three, one does not obtain this logical finding if one assumes aircraft valuation is constant over time, i.e., an annual valuation decline rate equal to zero.

These estimated PDM acceleration values can be applied to a cohort of aircraft. For instance, in FY 2005, WR completed PDM on 101 aircraft. We estimate that 33 were on their second PDM visit, 30 on their third, and 38 on their fourth. If all 101 aircraft had completed PDM 30 days faster, the estimated net benefit would have been about $10.7 million with a 1.35-percent valuation decline rate and $12.1 million with a 1.7-percent valuation decline rate (and $5.5 million with no valuation decline).

Next, we explore various extensions to our approach to assess the robustness of our findings.

Table 4.3
AFTOC FY 2005 F-15 Expenditures

Category	Subcategory	Expenditures (millions of FY 2005 $)
Ordinary operations	Mission personnel	1,142.9
	Unit-level consumption	1,650.7
	Intermediate maintenance	0.1
	Contractor support	64.9
	Sustaining support	56.2
	Indirect support	181.0
Subtotal		3,095.8
Depot maintenance		358.0
Total		3,453.8

Figure 4.4
Different Valuation Decline Curves

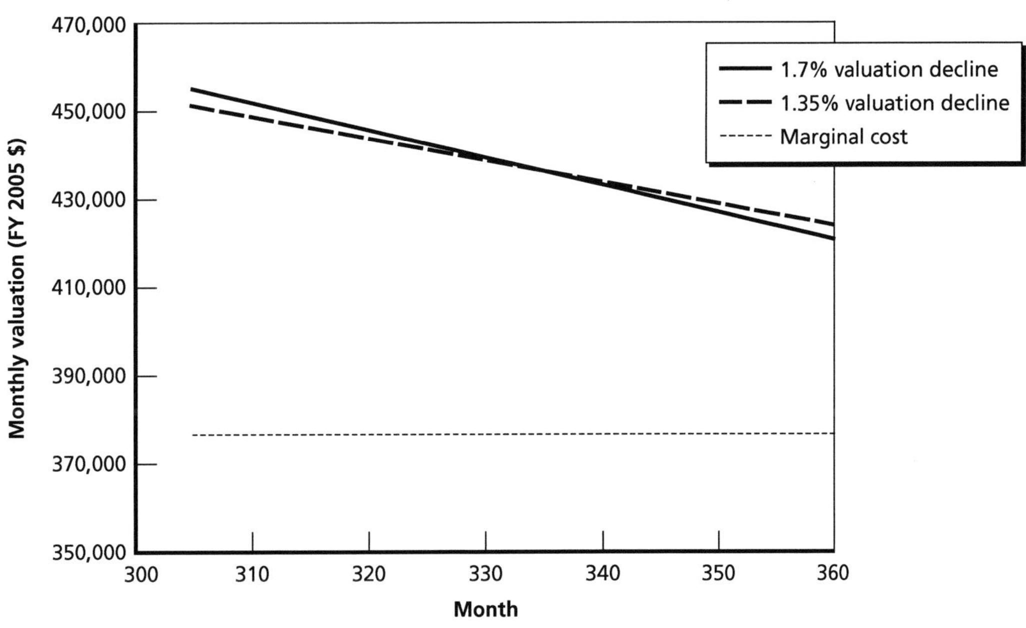

RAND *TR377-4.4*

Figure 4.5
Different Valuation Decline Curves Over F-15 Lifespan

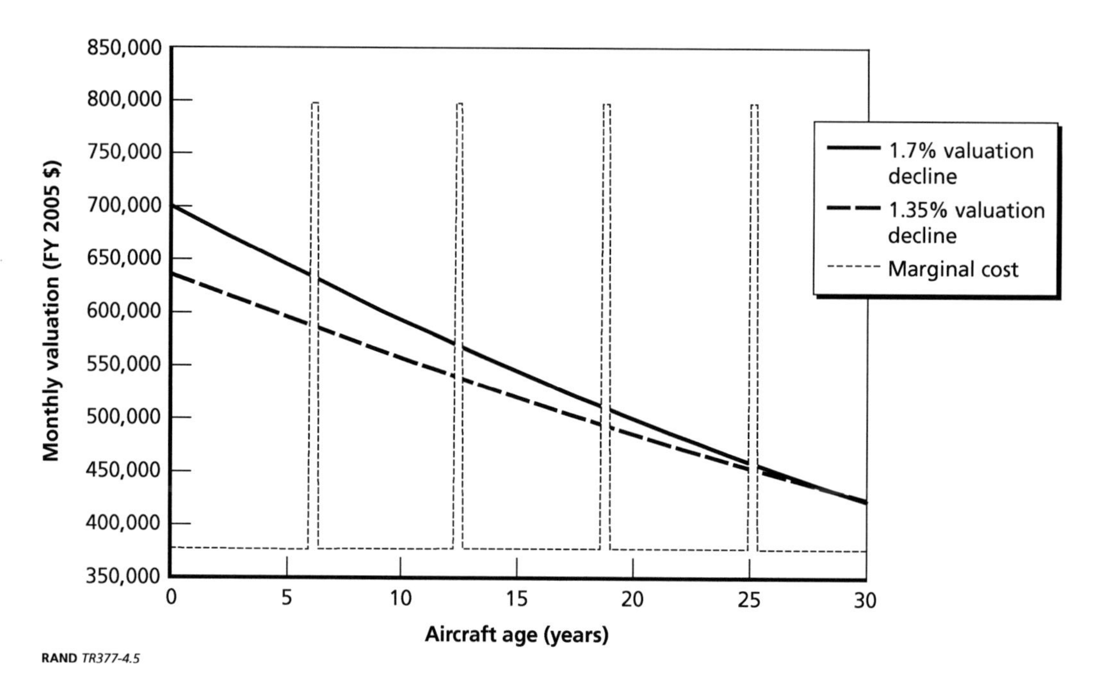

RAND *TR377-4.5*

Figure 4.6
Estimates of the Value of Accelerating Different PDM Visits

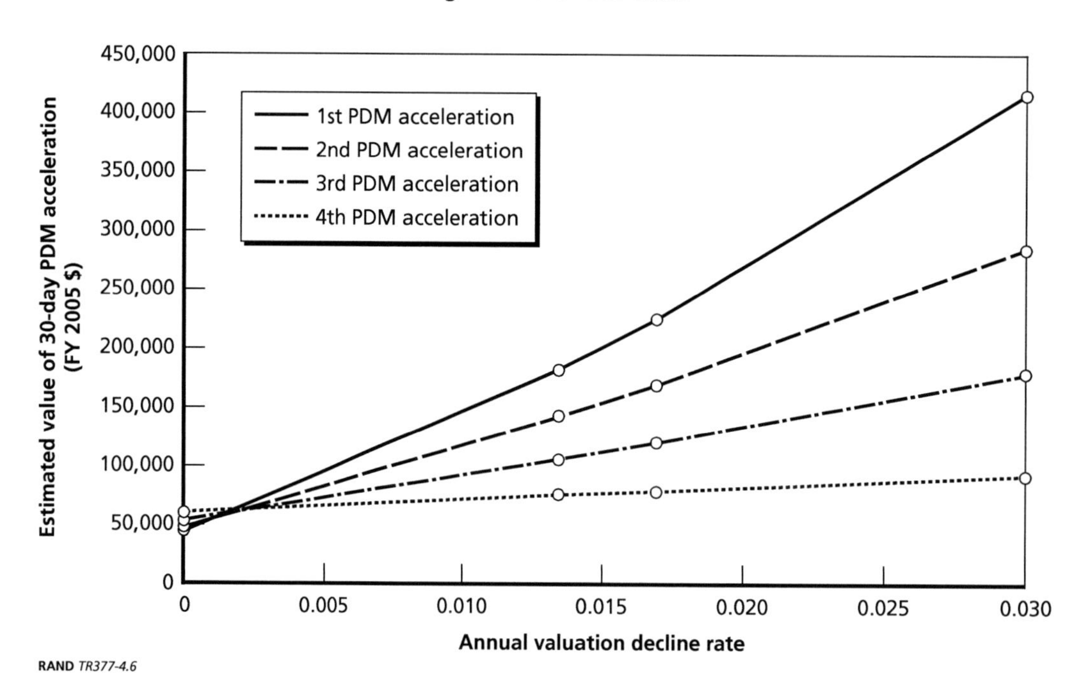

RAND *TR377-4.6*

CHAPTER FIVE

Robustness Explorations

In this chapter, we extend our inquiry in three directions:

- aging aircraft whose operating and PDM costs change over their lifetimes
- a constraint that expected total benefits over an aircraft's lifetime equal or exceed expected total costs
- the possibility of post-PDM valuation jumps and the impact of such jumps on our calculations.

The aging aircraft and life-cycle net benefit extensions are consistent in the sense that both extensions drive up (or at least do not decrease) our estimate of the value of accelerated PDM. Consideration of post-PDM valuation jumps does not change our valuation of accelerating an aircraft's last PDM visit, but it does potentially reduce the value of accelerating a PDM visit earlier in an aircraft's life.

Aging Aircraft

In Figure 4.5, we assume a constant $376,000 ordinary operations cost for each operating command–possessed month over the aircraft's life. By contrast, Pyles (2003) suggests that there is a "learning-effect period" in which initially high on-equipment maintenance workloads decline over the first five years of an aircraft's life. Then there is a period of considerable on-equipment maintenance growth when the aircraft is between the ages of five and ten, with a slower rate of growth thereafter.

Figure 4.5 also assumes a constant $3.2 million cost per PDM. Pyles found a pronounced acceleration of PDM workload as fleets age. (His greatest acceleration, however, occurs past our assumed 30-year F-15 retirement age.)

Variable ordinary operations and PDM costs can be handled by our valuation approach. Suppose, for instance, ordinary operations costs decline at a 2.5-percent real annual rate for the first five years of ownership[1] then ascend at a 1.5-percent real rate during years five through ten, then at a 1-percent rate thereafter. Suppose, too, the last PDM (when the aircraft is 25 years old) is 30 percent more expensive, in real dollars, than the earlier PDM visits. Instead of each PDM visit costing about $3.2 million, we now assume the first three visits each cost about $2.85 million while the last visit costs about $3.7 million. (These parameter estimates are purely illustrative, except in that they are intended to be broadly consistent with Pyles' findings.)

Figure 5.1 juxtaposes our baseline (Figure 4.5) and aging aircraft extension incremental cost assumptions. The aging aircraft case has rising ordinary operations costs past age five, plus its fourth and last PDM visit is more expensive than the other three. (We have also adjusted the first month's cost so that the discounted life-cycle costs of the two cases are the same.) With this new aging aircraft cost structure, we can repeat our estimation of PDM acceleration valuation.

In Figure 5.2, we reprise the computation in Figure 4.5, but using our new aging aircraft cost structure. Incorporation of aging aircraft maintenance cost effects consistently raises our estimated value of PDM acceleration. Figure 5.3 compares the baseline and aging aircraft extension estimates of the value of accelerating the fourth PDM visit by 30 days.

Figure 5.1
Aging Aircraft Extension Versus Baseline Assumed Monthly Incremental Costs

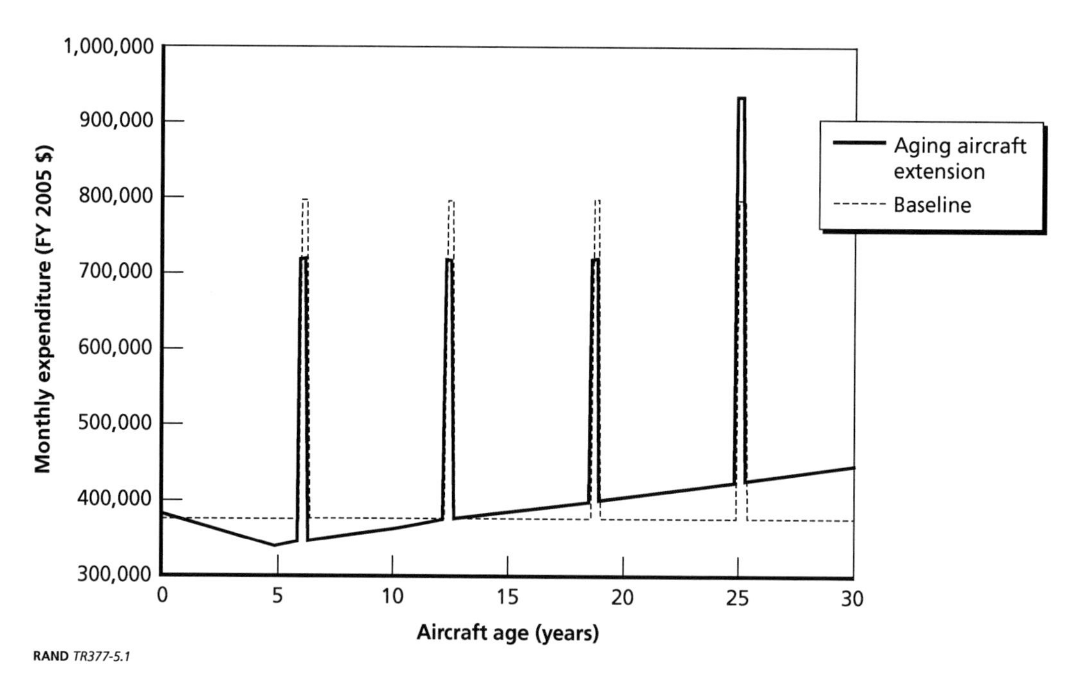

RAND *TR377-5.1*

[1] Such an early years' decline in monthly cost, C_m, could violate our assumption in Chapter Three that $B_m - C_m$ is nonincreasing. Loosening this assumption could theoretically result in a PDM constraint other than the binding "fourth PDM is worthwhile" constraint. However, with the parameters we use, we find the other three PDM visits are comfortably worthwhile if the fourth PDM is worth undertaking

Figure 5.2
Aging Aircraft Extension Estimated Monthly Costs and Benefits of F-15 Ownership

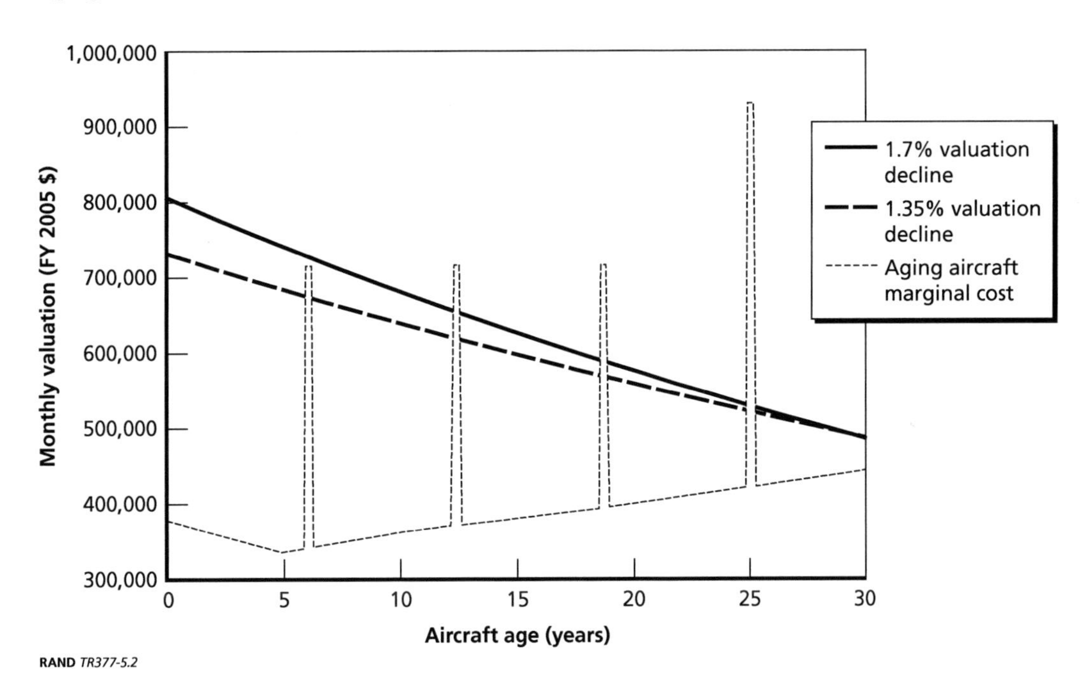

RAND *TR377-5.2*

Figure 5.3
Estimated Valuation of Accelerating the Fourth PDM Visit by 30 Days

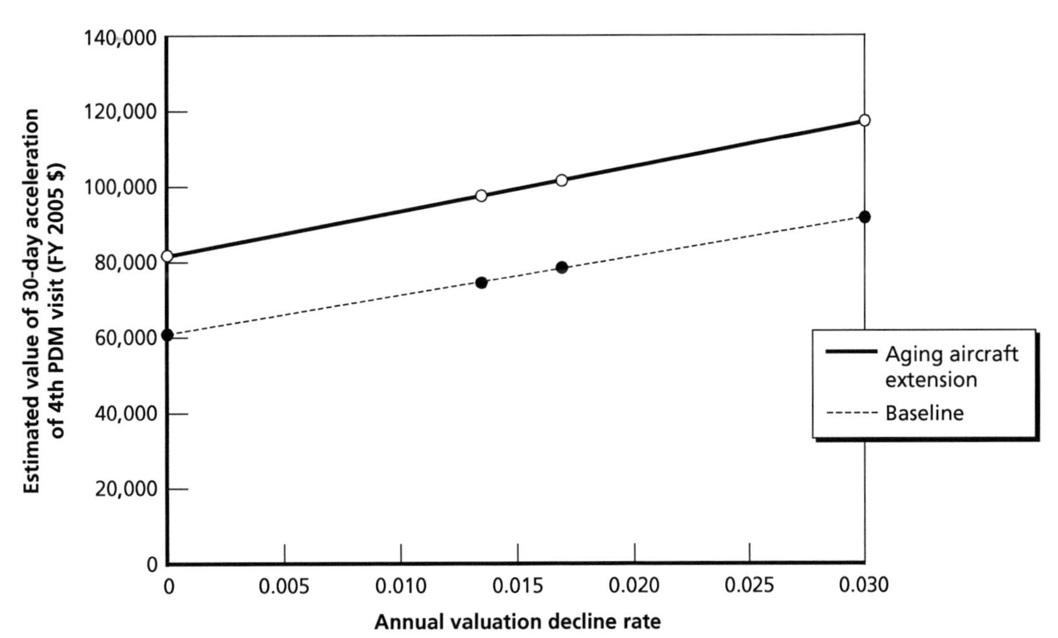

RAND *TR377-5.3*

There are two reasons that acceleration value estimates increase when aging aircraft phenomena are considered. First, when one assumes that the fourth PDM visit is more costly ($3.7 million rather than $3.2 million), one raises the level of surplus that must be obtained after that visit (assuming that undertaking the last PDM visit was appropriate). Second, with rising marginal costs, month 304 has a lower estimated marginal cost (about $422,000) than months 305–360 (which average about $432,000) so surplus from adding the extra month is greater than if marginal costs are constant.

If the Air Force faces aging aircraft challenges, it can respond in two ways. It can try to keep spending constant, but allow performance and availability to degrade. (Figure 4.5 is consistent with this approach.) Alternatively, it can spend enough to keep performance and availability constant (though the downward drift in Figures 4.2 and 4.3 is inconsistent with this supposition). Or, as is most likely what is happening, the Air Force can follow a mixed strategy of increasing spending somewhat while simultaneously allowing aircraft availability to degrade to an extent.

This exploration of how aging aircraft effects potentially increase PDM and ordinary operations costs reinforces our assertion that the previous chapters' estimates of the value of accelerated PDM are conservative lower bounds.

Consideration of Aircraft Acquisition Costs

Our analysis heretofore has used only PDM and ordinary operations costs to infer the monetary value of benefits generated by an F-15. Information about aircraft acquisition costs can give us additional insights into the value of benefits anticipated at the time the Air Force acquired the F-15. The Air Force indicates the acquisition cost of an F-15C/D was about $29.9 million in FY 1998 dollars[2] or about $34.6 million in FY 2005 dollars.

Aircraft acquisition costs, while clearly sunk, would be relevant if one believed that aircraft valuation, over its life cycle, must be high enough to not only cover the costs of PDM visits but all the system's costs:

$$\sum_{m=1}^{360} \frac{(B_m - C_m)}{(1+d)^{(\frac{m-1}{12}+\frac{1}{24})}} \geq \$34.6\text{M}$$

where the C_m values also include PDM costs. Otherwise, the Air Force should not have purchased the aircraft in the first place.

Aircraft acquisition costs are especially important in the constant valuation scenario in Chapter Three. If aircraft valuation is assumed to be constant (in real terms) over time, instead of requiring $61,164 in constant monthly surplus to justify PDM costs, one needs $187,321 in constant monthly surplus to additionally justify aircraft acquisition costs. See Figure 5.4.

---

[2] See U.S. Air Force (2005).

Figure 5.4
Different Constant Valuations with Different Constraints Imposed

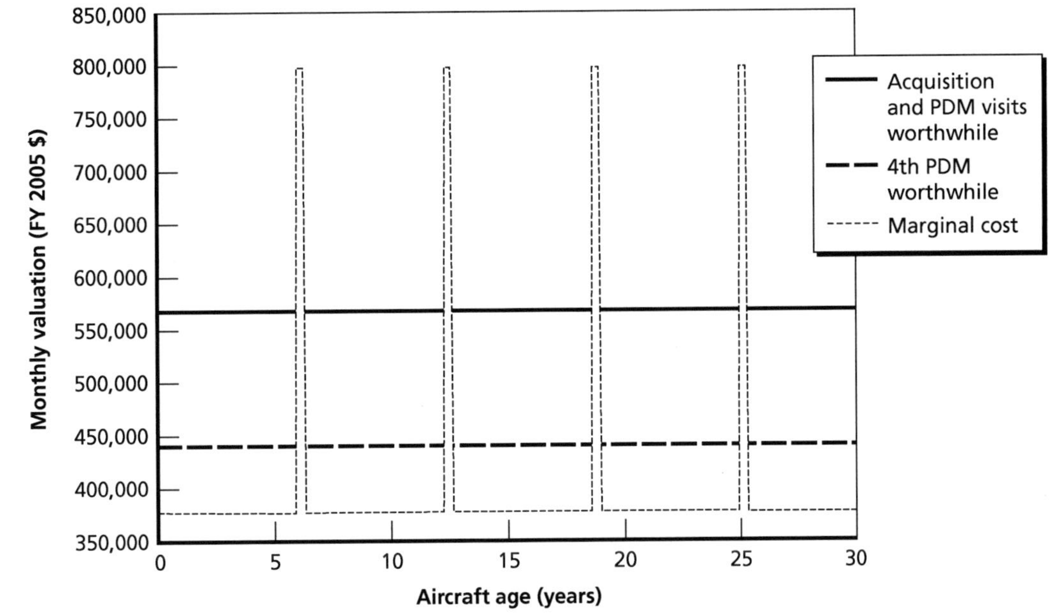

RAND *TR377-5.4*

The estimated value of expediting the fourth PDM visit by a month jumps from \$60,639 to \$185,713.

Intriguingly, when a positive valuation decline rate is considered, acquisition costs become less relevant. Indeed, if the valuation decline rate is great enough, a constraint that total life-cycle surplus justifies the acquisition cost is not binding. With a downward sloping valuation curve, having enough surplus to justify the final PDM visit,

$$\sum_{m=305}^{360} \frac{(B_m - C_m)}{(1+d)^{(\frac{m-301}{12}+\frac{1}{24})}} \geq \$3.2\text{M}$$

often implies a sizable surplus in earlier years of operation.

In Figure 5.5, we show the minimum valuation curves needed to justify the last PDM and acquisition cost constraints with a 1.35-percent valuation decline rate.

With a 1.35-percent valuation decline rate, fulfilling the last PDM constraint implies a minimum value of \$74,366 to accelerate the last PDM versus \$95,620 when the acquisition cost constraint is also applied.

With a 1.7-percent valuation decline rate, the acquisition cost constraint is no longer binding. Accelerating the fourth PDM visit has an estimated value of \$77,969 irrespective of whether the acquisition cost constraint is applied.

Curiously, the estimated value of accelerating PDM is not monotonically increasing in the valuation decline rate if both the PDM being worthwhile and acquisition cost constraints are imposed. See Figure 5.6.

Figure 5.5
Different 1.35-Percent Declining Valuations with Different Constraints Imposed

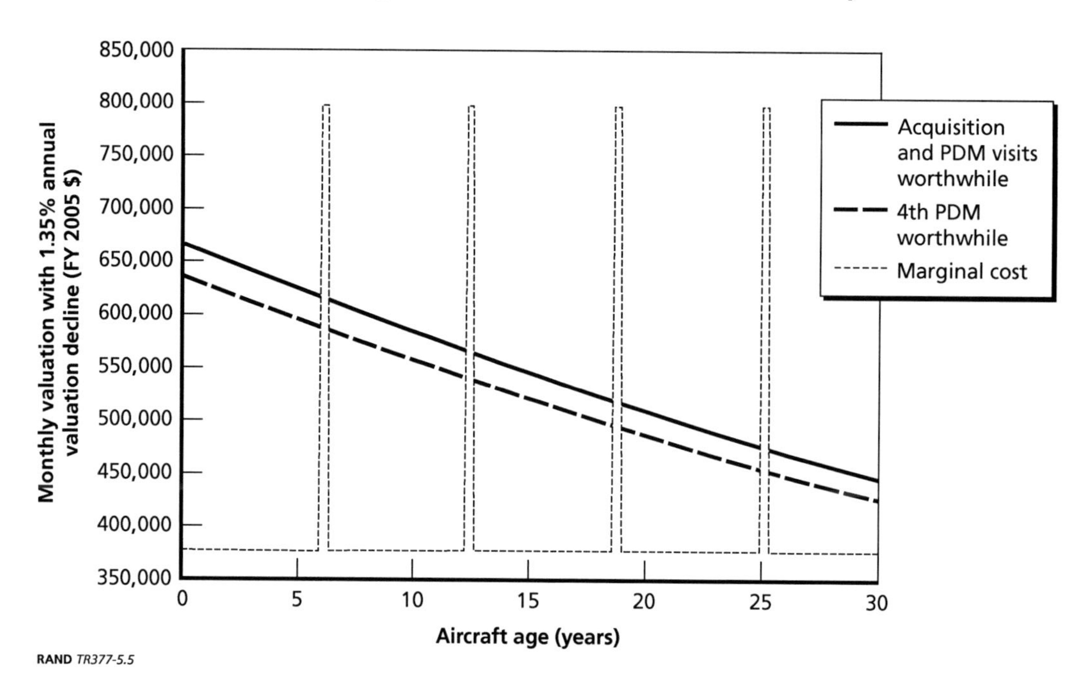

RAND *TR377-5.5*

Figure 5.6
Estimated Value of Accelerating PDM with PDM and Acquisition Cost Constraints Imposed

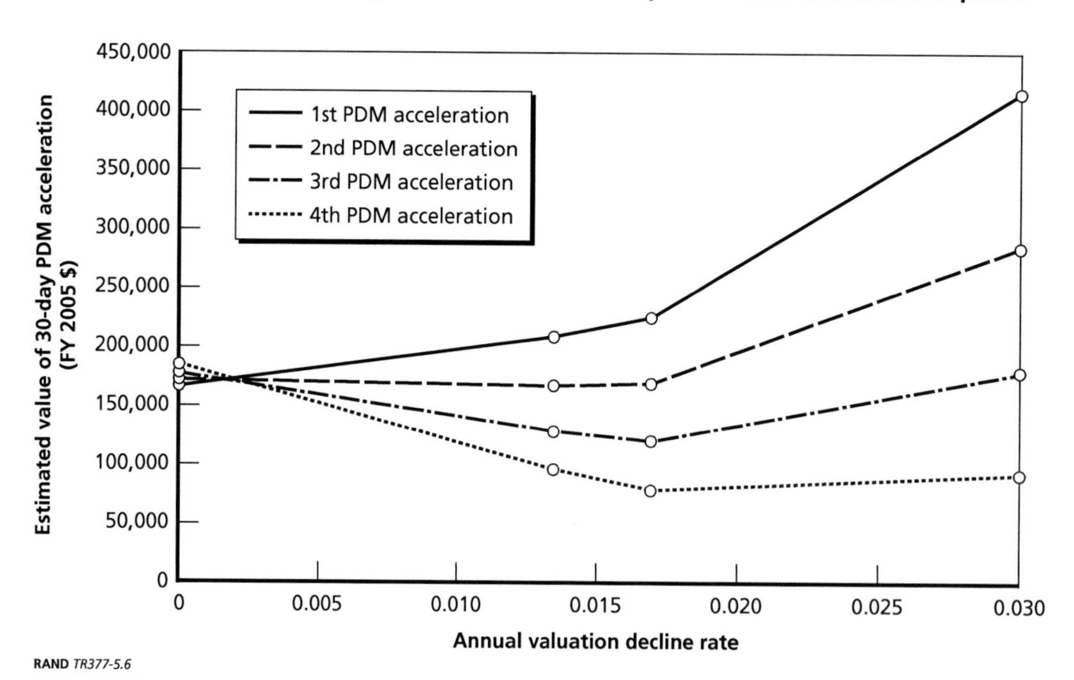

RAND *TR377-5.6*

For low valuation decline rates, the acquisition cost constraint binds and valuation must be quite high after the fourth PDM. For higher valuation decline rates, the fourth PDM constraint binds and acquisition costs are easily justified.

We are not sure of the appropriateness of the acquisition cost constraint. Most of the F-15 fleet was acquired during the Cold War, so one might argue that an investment that was viewed as appropriate in the 1980s may not have similar value today. The fourth PDM constraint, by contrast, is much more justifiable, as the Air Force could, with cognizance of the current security environment, decide to retire an aircraft rather than undertake a PDM visit.

Imposing the acquisition cost constraint never lowers the estimated value of PDM acceleration, again supporting our view that our approach imposes a lower bound on what PDM acceleration might be worth.

Post-PDM Valuation Jumps

One might think that putting an aircraft through PDM makes the aircraft more valuable, e.g., it is more reliable or has received capability improvements. (If this were so, we might have operating command-possessed B_m jumps, e.g., $B_{305} > B_{300}$.)

As discussed in Chapter Four, one metric that may be correlated with aircraft valuation is an aircraft's mission capability rate. We therefore assessed whether going through PDM increases mission capability rates.

We identified 89 F-15Cs and F-15Ds that entered WR PDM after October 1, 2002, and were completed by September 30, 2004. We then computed each aircraft's average MC and FMC rates in the full 12 months preceding each aircraft's entering PDM and the full 12 months following PDM. We disregarded the months in which the aircraft entered and left WR as well as the months spent at WR. We chose our sample based on the criterion that we wanted at least 12 months of observations after the beginning of FY 2002 but before PDM commenced and at least 12 months of observations after PDM completion.

In this sample, the average post-PDM MC rate was slightly higher than the pre-PDM rate (78.0 percent versus 76.7 percent), as was the average post-PDM FMC rate (76.9 percent versus 75.0 percent). For neither metric, however, does single-factor analysis of variance find the difference to be statistically significant ($p = 0.32$ for MC, $p = 0.20$ for FMC).

Obviously, the post-PDM aircraft were older. Given the secular trend for diminishing MC and FMC rates and the fact that post-PDM aircraft had greater, not lower, MC and FMC rates, this is consistent with (though does not prove) the assumption that aircraft become more valuable as a result of PDM visits.

To understand how a positive jump in benefit following PDM would affect our approach, we revisited the nonaging model in Chapter Four, adding a 5-percent real increase in aircraft valuation following a PDM visit. Figure 5.7 presents our findings.

Figure 5.7
Valuation Curves with Post-PDM Valuation Jumps

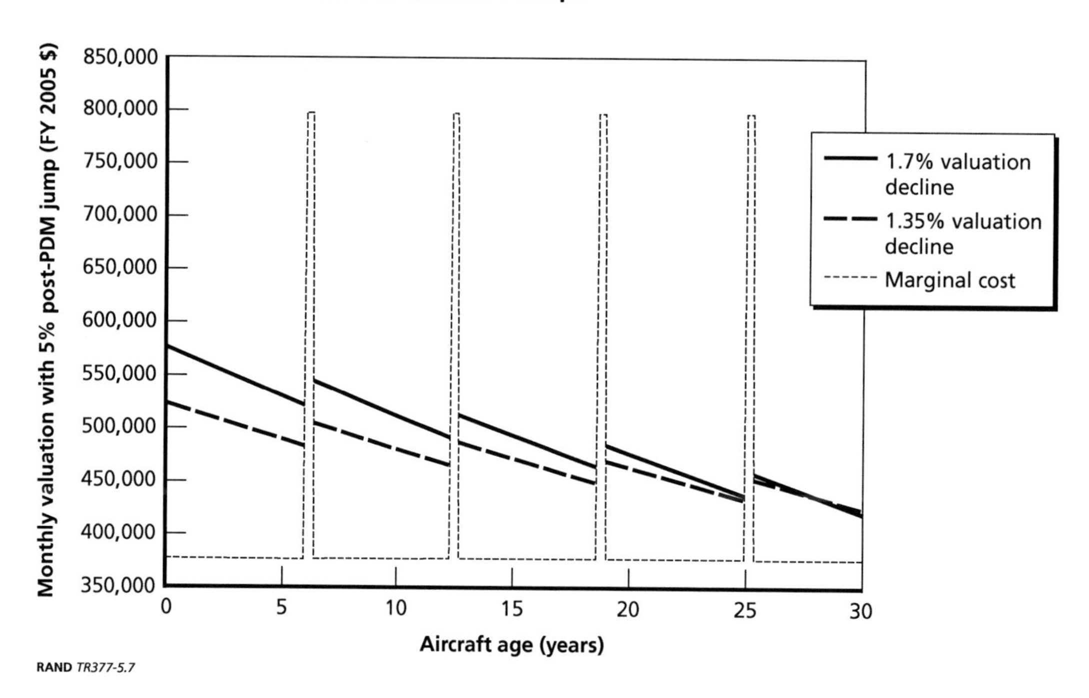

RAND *TR377-5.7*

The sawtooth shape emanates from the PDM valuation increase: The valuation curve jumps up after each PDM visit.

A valuation-jump structure of this sort has no effect on the minimum valuation associated with accelerating the last PDM visit. We have already derived the minimum possible value of accelerating the last PDM visit consistent with that last visit being worth undertaking.

Instead, the effect of the post-PDM valuation jumps is to make the minimum value of accelerating an earlier visit lower. Valuation can be lower in earlier years because of the post-PDM jumps. In Figure 5.8, we use the 1.35-percent valuation decline case and show how our estimates of the value of accelerating earlier PDM visits decline as the estimated post-PDM valuation jump increases.

Post-PDM valuation jumps are irrelevant in our model as it pertains to estimating the value of accelerating the last PDM visit. The presence of post-PDM jumps reduces the value of accelerating earlier PDM visits.

While it is plausible that there are post-PDM valuation jumps, we did not find any statistically significant evidence that aircraft have increased mission capability rates following PDM.

Figure 5.8
The Effect of Post-PDM Jumps on PDM Acceleration Valuation

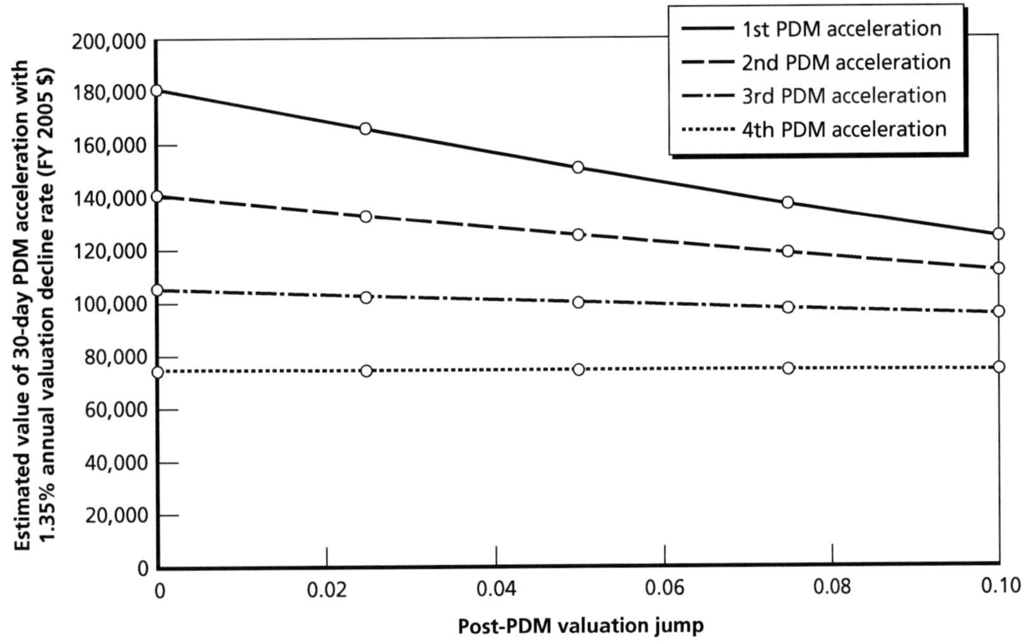

RAND *TR377-5.8*

CHAPTER SIX

Conclusions

The goal of this report was to provide Warner Robins and other Air Force decisionmakers with a heuristic as to how much expediting F-15 programmed depot maintenance might be worth.

Our valuation technique is built around revealed preference. In particular, we assume an F-15 must have enough value after PDM to justify the cost of the PDM visit. Otherwise, the aircraft should have been retired. This is an inherently conservative assumption. Perhaps aircraft are worth far more to the Air Force than what it pays for PDM.

Given an aircraft valuation curve consistent with PDM being worthwhile, we can estimate what an extra month of operating command possession emanating from expedited PDM would be worth.

If we assume that aircraft valuation does not increase with age, expediting an F-15's last PDM visit by one month must be worth at least $60,000. This value emanates from a horizontal valuation line wherein an aircraft's real valuation is constant over its lifetime. With a horizontal valuation line, it is least valuable to expedite a relatively new F-15 undergoing PDM for the first time.

We do not think that a horizontal valuation line is plausible. We expect aircraft valuations to decline as aircraft age: Rivals' technology improves and aircraft suffer wear-and-tear.

An analysis of 11 years of F-15C/D data finds a secular trend of declining mission capable and fully mission capable rates. Declining MC and FMC rates may cause the Air Force's valuation of the F-15 to decline or may be symptomatic of such a decline.

When declining aircraft valuation over time is considered in our structure, our estimates of the value of expediting PDM increase (see Table 6.1). Declining aircraft valuation also implies that it is more valuable to expedite a new F-15's first PDM visit than an old F-15's last PDM visit.

Table 6.1 also shows the results of robustness explorations. Consideration of aging aircraft issues (e.g., older F-15s' PDM visits are more expensive) increases our estimates of the value of expediting PDM.

One can also add a constraint that an aircraft's life-cycle benefits are sufficiently large to justify the aircraft's acquisition costs (not only its PDM costs). This additional constraint increases our estimated value of expedited PDM if the valuation decline rate is low, but it is irrelevant if the valuation decline rate is high enough.

Table 6.1
Different Estimates of the Value of Expediting a PDM Visit by One Month

Valuation Case	Robustness Case	Estimated Value by PDM Visit ($)			
		1st	2nd	3rd	4th
Constant	4th PDM visit worthwhile	43,077	47,875	53,661	60,639
1.35% annual decline	4th PDM visit worthwhile	181,639	141,232	105,508	74,366
	Aging aircraft	297,917	222,276	156,614	96,489
	Acquisition cost	208,396	165,959	128,404	95,620
	5% post-PDM jump	150,763	125,615	100,225	74,366
1.7% annual decline	4th PDM worthwhile	224,543	168,661	119,946	77,969
	Aging aircraft	347,278	253,833	173,224	100,634
	Acquisition cost	224,543	168,661	119,946	77,969
	5% post-PDM jump	189,356	151,154	114,117	77,969

NOTE: Estimated values are in FY 2005 dollars.

Post-PDM valuation jumps do not affect our estimate of the value of expediting an F-15's last PDM visit but decrease the (still greater) value of expediting earlier PDM visits.

While this report has focused on the F-15, our approach could be applied to any system that periodically enters depot-level maintenance. A lower bound on the value of expediting PDM can be estimated from the assumption that net benefits following PDM equal or exceed the costs of PDM.

APPENDIX A

Calculating PDM Acceleration Valuations

This appendix presents how our acceleration valuations were computed using Microsoft Excel. In particular, we work through the case of the 1.7-percent valuation decline curve in Figure 4.5 and show how valuations are computed for accelerating different PDM visits.

We assume that the aircraft's marginal cost (C_m) in months 305–360 is \$376,369. We also assume PDM in months 301–304 costs \$795,923 per month. Measured from the beginning of month 301, with a 3-percent real discount rate, the present value of the sum of future expenditures is \$22,662,640:

$$\sum_{m=301}^{360} \frac{C_m}{(1+d)^{(\frac{m-301}{12}+\frac{1}{24})}} \text{ with } d = 0.03.$$

Benefits from having the aircraft possessed by operators (B_m) are to accrue in months 305–360 (without PDM acceleration). We assume that aircraft accrue no benefits while in PDM. We assume that the present value of the sum of future benefits, measured from the beginning of month 301 when the PDM visit commences,

$$\sum_{m=305}^{360} \frac{B_m}{(1+d)^{(\frac{m-301}{12}+\frac{1}{24})}},$$

must at least equal \$22,662,640.

We are further assuming that aircraft valuation decreases at a 1.7-percent annual rate. We have disaggregated time into monthly blocks, so, incorporating benefit decline, we impose that $B_{359} = (1+0.017/12) * B_{360}$, $B_{358} = (1+0.017/12) * B_{359}$, and so forth up to $B_{305} = (1+0.017/12) * B_{306}$. In Excel terms, we have $P350 = +P351 * (1+0.017/12)$, for example.

The one unknown in this calculation is B_{360}; all other months' values are determined from month 360.

We created the discounted sum of benefits accruing from months 305 through 360, measured as of the beginning of month 301. We then used Excel's goal-seek capability (**Tools | Goal Seek**) to find the value of the month 360 benefit that sets the discounted benefit sum equal to $22,662,640. This value turned out to be $420,334, implying that the month 305 estimated benefit was $454,370.

If the last PDM visit were completed a month sooner, benefit would accrue in month 304. Month 304 would have a benefit of 1 + 0.017/12 times the month 305 level, or $455,013. Subtracting an additional operating month's marginal cost of $376,369 leaves a net benefit of $78,644 in month 304. But then we wanted the net benefit computed as of the beginning of month 301 (when the PDM visit commences) so it totaled $77,969 with appropriate discounting:

$$\frac{78{,}644}{1.03^{\frac{3.5}{12}}}.$$

Once the monthly benefit curve with $B_{360} = \$420{,}334$ is derived, the benefit curve can be computed back to month 1. We do this by repeatedly applying the formula $B_m = (1 + 0.017/12) * B_{m+1}$. Our estimated benefit in the first month of operating command possession is $698,736. Months to be spent in PDM, however, have their benefit set equal to zero.

With the life-cycle benefit curve in place, it is fairly straightforward to calculate the benefit of accelerating earlier PDM visits. The gross nominal benefit for month 228 (the fourth month of the third PDM visit) would be $506,699 if it were no longer a PDM month. The net benefit measured at the beginning of month 225 (when the third PDM visit commences) is $129,211. The added wrinkle in expediting the third PDM visit is that one brings forward the fourth visit. Measured from month 225, losing operating command possession in month 300 costs $67,443, while gaining it in month 304 accrues $64,657. (We assume, after one-time PDM acceleration, that it returns to a 120-day calendar thereafter.) Additionally, the fourth PDM visit's $3.2 million cost shifts forward by a month. This shift, measured from month 225, costs $6,479, so the net benefit of accelerating the third PDM visit by one month is estimated to be worth $119,946 ($\$129{,}211 - \$67{,}443 + \$64{,}657 - \$6{,}479$). Similar procedures (with multiple shifts forward) apply to computing the values of accelerating the second and first PDM visits.

PDM Speed for New Military Aircraft

In theory, the speed of PDM could influence the initial acquisition of an aircraft. Specifically, if a lengthy PDM duration were forecast, the Air Force would need a larger fleet to assure that a requisite number of aircraft is available. Conversely, quicker PDM would allow the purchase of a smaller fleet. It may be worthwhile to invest in expedited PDM if it allows the acquisition of fewer aircraft.

Suppose the Air Force acquires N aircraft. Suppose (as is true of the F-15) that these aircraft are on a six-year PDM cycle with each PDM visit lasting D days. The steady-state number of aircraft available to operating commands (i.e., not in PDM) would be

$$\frac{N * 365.25 * 6}{365.25 * 6 + D} = \frac{2{,}191.5 * N}{2{,}191.5 + D}.$$

(As of September 30, 2005, there were 722 F-15s, so if $D = 120$, the steady-state number of aircraft available to operators would be about 684.5. Table 2.2 shows that 684 F-15s were possessed by operating commands on that date.)

Suppose there is an exogenous constraint that the steady-state number of operating command–possessed aircraft must be at least k. Then

$$k \leq \frac{2{,}191.5 * N}{2{,}191.5 + D},$$

so

$$N \geq \frac{k * (2{,}191.5 + D)}{2{,}191.5} = k + \frac{D * k}{2{,}191.5}.$$

For each additional required steady-state operating command–possessed aircraft, the Air Force must acquire

$$1 + \frac{D}{2{,}191.5}$$

additional aircraft. If D increases by one, the Air Force must acquire

$$\frac{k}{2{,}191.5}$$

additional aircraft.

If the Air Force has estimated the life-cycle cost of an aircraft, as well as the costs of changing PDM speed, it can determine the cost-minimizing levels of N and D.

Define C to be the discounted life-cycle (acquisition and support) costs of an airplane, excepting those costs related to the choice of D. Define the total life-cycle costs of an aircraft to be $C + f(D)$. $f(D) > 0$ but

$$\frac{\partial f(D)}{\partial D} < 0,$$

i.e., it is more costly to speed up PDM.

If the Air Force acquires N aircraft, its total costs would be $N * C + N * f(D)$. The Air Force wants to choose the cost-minimizing level of D subject to the constraint

$$k \leq \frac{2{,}191.5 * N}{2{,}191.5 + D}$$

or

$$N \geq \frac{k * (2{,}191.5 + D)}{2{,}191.5}.$$

Substituting in the constraint, the problem becomes choosing D to minimize

$$\frac{C * k * (2{,}191.5 + D)}{2{,}191.5} + \frac{f(D) * k * (2{,}191.5 + D)}{2{,}191.5}.$$

Eliminating irrelevant terms, the minimand is $C * D + 2{,}191.5 * f(D) + f(D) * D$. Differentiating with respect to D, we get

$$C + 2{,}191.5 * \frac{\partial f(D)}{\partial D} + \frac{\partial f(D)}{\partial D} * D + f(D) = 0.$$

This equation does not have a general solution. For illustrative purposes, however, suppose $f(D)$ has the functional form

$$f(D) = \frac{g}{D},$$

so

$$\frac{\partial f(D)}{\partial D} = -\frac{g}{D^2}.$$

Our equation then becomes

$$C - \frac{2{,}191.5 * g}{D^2} - \frac{g}{D} + \frac{g}{D} = 0,$$

so the optimal

$$D^* = \sqrt{\frac{2{,}191.5 * g}{C}}$$

and

$$N^* = k + k * \sqrt{\frac{g}{2{,}191.5 * C}}.^{1}$$

As C increases relative to g (i.e., PDM speed costs become less important), the optimal D^* falls, as does the optimal N^*. The Air Force will want costlier aircraft to go through PDM more quickly so fewer aircraft can be purchased up front, other things being equal.

Suppose, an F-15's life-cycle cost was in the form of

$$C + f(D) = 125{,}000{,}000 + \frac{444{,}000{,}000}{D},$$

and suppose the minimum number of acceptable available aircraft was $k = 680$. Then

$$D^* = \sqrt{\frac{2{,}191.5 * g}{C}} = 88$$

[1] Having a closed-form, analytic solution for D^* and N^* is the exception, not the rule. For instance, if

$$f(D) = \frac{h}{D^2},$$

the first-order condition for the optimal value of D is the cubic equation $CD^3 - hD - 4{,}383h = 0$.

and

$$N^* = k + k * \sqrt{\frac{g}{2{,}191.5 * C}} = 708.$$

References

Keating, Edward G., and Frank Camm, *How Should the U.S. Air Force Depot Maintenance Activity Group Be Funded? Insights From Expenditure and Flying Hour Data*, Santa Monica, Calif.: RAND Corporation, MR-1487-AF, 2002. Online at http://www.rand.org/pubs/monograph_reports/MR1487/ (as of May 15, 2006).

Keating, Edward G., and Matthew Dixon, *Investigating Optimal Replacement of Aging Air Force Systems*, Santa Monica, Calif.: RAND Corporation, MR-1763-AF, 2003. Online at http://www.rand.org/pubs/monograph_reports/MR1763/ (as of May 15, 2006).

Keating, Edward G., Don Snyder, Matthew Dixon, and Elvira N. Loredo, *Aging Aircraft Repair-Replacement Decisions with Depot-Level Capacity as a Policy Choice Variable*, Santa Monica, Calif.: RAND Corporation, MG-241-AF, 2005. Online at http://www.rand.org/pubs/monographs/MG241/ (as of May 15, 2006).

Office of Management and Budget, *Circular No. A-94, Appendix C: Discount Rates for Cost-Effectiveness, Lease Purchase, and Related Analyses*, January 2006. Online at http://www.whitehouse.gov/omb/circulars/a094/a94_appx-c.html (as of February 14, 2006).

Pyles, Raymond A., *Aging Aircraft: USAF Workload and Material Consumption Life Cycle Patterns*, Santa Monica, Calif.: RAND Corporation, MR-1641-AF, 2003. Online at http://www.rand.org/pubs/monograph_reports/MR1641/ (as of May 15, 2006).

Reinertsen, Donald G., Ross T. McNutt, Michael A. Greiner, and Frank E. Hutchison, "An Overview of Cost-of-Delay Analysis: Calculating Project Decision Rules," *The Journal of Cost Analysis and Management*, Winter 2002, pp. 8–24.

U.S. Air Force, "F-15 Eagle," fact sheet, Langley Air Force Base, Va.: Air Combat Command, Public Affairs Office, October 2005. Online at http://www.af.mil/factsheets/factsheet.asp?fsID=101 (as of January 20, 2006).

Younossi, Obaid, Mark V. Arena, Michael Boito, Jim Dryden, and Jerry Sollinger, *The Eyes of the Fleet: An Analysis of the E-2C Aircraft Acquisition Options*, Santa Monica, Calif.: RAND Corporation, MR-1517-NAVY, 2002. Online at http://www.rand.org/pubs/monograph_reports/MR1517/ (as of May 15, 2006).